THE NE

Wheels Around EDITION

Carole Wilkinson

A word from our editor

ALL over the world, by day and by night, millions of wheels turn around and around. They carry us and they transport our food.

There are other wheels that don't travel anywhere. They spin on the spot, making machines work. They even help us tell the time.

In this edition of *The News*, we discover the secrets behind wheels.

We find out where wheels can be found and why they are so useful.

Wonderful wheels—where would we be without them? We'd be standing still!

Daily spot

Aeroplane wheels have to be pumped up six times harder than car tyres. They are filled with **nitrogen**, not air.

INSIDE

Why were wheels invented?

Moving with the times.

WHEELS WERE invented because people needed to be able to move large, heavy objects. It takes a lot of energy to move something. It has to be pushed, pulled or lifted. The energy it takes to move things is called **force**.

People didn't like having to use so much force. They didn't like all that pushing, pulling or lifting. They got tired and sweaty. It was hard work. With wheels, life suddenly got a whole lot easier.

Putting the box on wheels makes it 100 times easier for the boy to move the box.

If he used smooth metal wheels on steel railway tracks, the box would be 300 times easier to move.

Getting the dog to pull it for him is the easiest of all. He doesn't have to do anything!

Fact of the day

Wheels can support about 400 times their own weight. They are some of the strongest things humans make.

Speedy wheels

The travel revolution.

THE FIRST thing that wheels did for people was to take the weight off their shoulders. They no longer had to struggle to carry heavy loads.

A horse and cart was very slow, though. Sometimes, people on foot overtook them!

Then, people got better at making wheels and building roads. They discovered that wheels travelled better on smoother roads. People soon found that wheels had another advantage—speed.

Engines were invented to turn the wheels. Cars, trucks and trains started going faster and faster.

With more speed, people could travel greater distances. Today, a trip that once took weeks can be completed in one day.

The News Cartoon

It's not easy to push a large load uphill but, thanks to wheels, these people don't have to carry the load on their backs!

Look for the road signs in the book. And watch for them on the roads.

The secret of wheels

Find out what a difference wheels can make.

TRY PUSHING something heavy along the ground. If it doesn't have wheels, it's hard work. Here's a way to find out how wheels make moving things easier.

You will need:

- a lunch box
- small heavy items, such as marbles, small rocks or potatoes
- six round pencils

What to do:

1. Fill the lunch box with the heavy objects.
2. Push it along the floor. Take note of how much force you need to push it.
3. Now get six round pencils and put them under the lunch box.
4. Gently push the lunch box again. The pencils act as wheels. You will find it is now much easier to move.

A fraction too much friction

What makes it hard to push something along the ground?

WHEN YOU push an object, such as the lunch box in our experiment, along a smooth floor, it makes a scraping sound. Even though you can't see them, there are tiny bumps and lumps on the object. There are also tiny lumps and bumps on the floor.

These lumps catch on each other and stop the lunch box from moving smoothly across the surface of the floor. This is called **friction**.

Friction is the force that slows things down when they move. A round object still has lumps and bumps. But as it rolls along the ground, the lumps and bumps don't get caught up with the lumps and bumps on the ground as much.

This means there is less friction and the object moves more easily.

Oil is used to reduce friction, so that the parts in cars and bicycles move more smoothly.

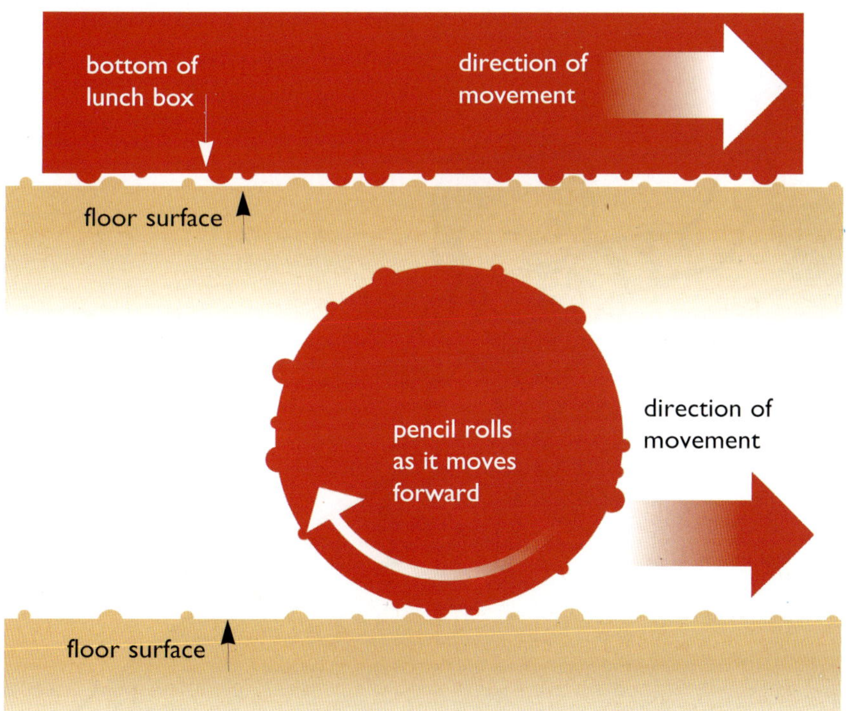

Did you know?

If you had two rocks that were exactly the same weight, but one was square and one was round, you would need five times more force to push the square rock along the ground.

Wheel-less world

Imagine... what it would be like in a world without wheels.

IF WHEELS hadn't been invented, how different would your life be?

Danny lives in a world without wheels. Let's take a look at a day in his life.

When Danny gets up in the morning, he's hungry. He opens the cupboard to see what's for breakfast. There are no cornflakes, no Vegemite. Without wheels, there are no trucks to transport food from farms and factories.

Danny has to make do with a slice of homemade bread and an apple from a tree in the back garden.

He pulls on his school clothes. They are made from animal skins and have been hand sewn by his mum. There are no spinning wheels, so cotton and wool can't be spun into thread to make cloth for clothes. There are no sewing machines either.

Danny checks the **sundial** clock on his wrist. It's time for school. He waves goodbye to his mum who is using a rock to grind grain to make flour to make bread.

Danny has to walk to school. Without wheels there are no bikes, no cars and no buses. Dad had to leave for work hours ago so he could ride his horse to the city.

At lunch time, Danny would probably love to play on a skateboard or roller-blades. He'll never know what he's missing, because they don't exist in his world.

There's not much room to play anyway. The space where the playground might have been is a vegetable garden and chicken yard. That's where the school grows the food to put in the sandwiches sold at the school canteen.

After school, Danny can't watch television because there's no electricity. (There are lots of wheels involved in making electricity.) Danny's family have a solar panel on the roof, but it's been cloudy. They have to save power for more important things.

Pedal pushing

People-power.

BICYCLES ARE very useful. They are small vehicles that don't have an engine to make them work.

Bicycles work with the power of a person's legs and feet. Even though they only have two wheels, they are easy to ride. Around the world, lots of children ride to and from school every day.

People all over the world use them. In some countries, a bicycle is like a family car. A man rides the bike. His wife sits on the pack rack with the baby strapped to her back. An older child balances on the crossbar.

A bicycle or a motor bike can also be like a moving truck. It can be used to carry many things. It could be the weekly shopping for a family, the luggage of a group of tourists or the goods for a shopkeeper's stall.

Riding to school with your friends is fun. But what if you had to balance a load like this? It's amazing the bike doesn't topple over!

Changing gears

Why is riding faster than walking?

HAVE A CLOSE look at the bike in the picture. There is a large **gear** wheel fitted to the pedals. The gear wheel is not smooth. It has lots of points on it. These points are called teeth. When the rider pushes the pedals, this gear wheel turns.

There is a smaller gear wheel with fewer teeth, attached to the back wheel. The bike chain fits into the teeth and runs around both wheels. Every time the little gear wheel turns, the back wheel moves with it.

The larger gear wheel is three times as big as the small one. This means that every time the big gear turns once, the little gear turns three times. Every time the pedals turn once, the bike wheels turn three times.

Riding a bike is quicker and easier than walking. It is a good use of our energy. A push of the pedal takes us much further than a footstep.

Wheels, wheels everywhere! See how many wheels you can find in this picture.

Did you know?

The very first bicycles were built in the early 1800s. They were made of wood with steel wheels and no pedals. To move them, riders had to use their feet to push off, like you do with a scooter!

Cars take over

Wheels go around the world.

MOST OF THE wheels in the world today are on cars. There are around 500 million cars in the world. That's a lot of wheels!

The wheels on cars are turned by an engine. The people who invented cars in the 1800s tried engines that ran on steam, electricity and petrol. They decided that engines that ran on petrol were the best. Cars have been running on petrol ever since.

Petrol is made from oil that is pumped up from **deposits** under the Earth. In the future, the oil will run out and there will be no more petrol. Before this happens, we will have to find another way to run cars.

Some people think that running cars on electricity is a good idea. Other people have been experimenting with cars that run on energy from the sun. These are called solar-powered cars.

Daily spot

In Los Angeles in the USA, there are 1,174 cars for every square km (3,040 per square mile).

Did you know?

When cars were first invented, some people didn't like them. They thought they were too noisy and smelly. They tried to stop them by putting nails and broken glass on the roads to **puncture** the tyres.

Axle Action

Adam Axle has broken his silence. He told our reporter about his part in the success of the wheel.

MR ADAM AXLE took time out from his busy schedule to talk to *The News*. He explained that an axle is a long thin rod made of wood or metal. It joins two wheels, such as the front or back two wheels of a cart or a car.

"We have kept quiet for long enough," said Adam Axle. "If it wasn't for axles, the only vehicles with wheels would be bikes!"

"People can't see us, so they don't know we exist. If they have heard of us, they never know how important we are," said the angry axle.

Mr Axle went on to say that since wheels were invented, more than five thousand years ago, his **ancestors** have been helping wheels.

"Right from the beginning, the earliest carts had an axle joining two wheels," he said.

Mr Axle, a steel axle, has been working with two wheels on the front of a tram for twenty years.

"It's time everyone knew the truth," Mr Axle said. "Without axles, wheels would be useless."

No wheels were available for comment.

Rolling along

What sort of wheels go round at the airport?

AT AN AIRPORT, you can see plenty of wings. There are also lots of wheels. They come in all types and sizes. Think about all the wheels you would see on your way to take a trip in a plane.

Just to get to the airport, you need wheels. These are the wheels on the car, bus or train that takes you there.

When you arrive, you have to get your luggage from the carpark to the check-in desk.

If there are wheels on your suitcase, you can push or pull it along. If you have lots of luggage, you can stack it all on a big trolley with wheels.

When all the passengers' luggage has been checked in, it's loaded onto a trailer and towed out to the aeroplane.

The wheels on the aeroplane are very important. They are used to **taxi** along the runway, and for take-off and landing.

For an aeroplane to land, the wheels must be down so that it can touch down smoothly and roll along the runway to the parking bay.

The strongest wheels

THE nose wheels on a 737 aircraft are about the same size as car wheels. But they have to be much stronger. Each wheel has to carry a weight of 16,000 kg (35,300 pounds).

The wheels hit the runway at about 125 kilometres (200 miles) per hour. Each tyre has to withstand an incredible force when the plane lands.

The tyres get very hot, about 750°C (1,382°F). That's more than three times hotter than an oven baking bread.

Aeroplane tyres are tested to make sure they won't explode until they get much hotter than they would on a runway.

The wheel facts

The wheel size of things.

The most wheels

Semi-trailers have 22 wheels each. A road train, which is a truck with two trailers attached, has 34 wheels.

The fewest wheels

A unicycle has just one wheel, and so does a wheelbarrow.

The biggest wheels

The biggest wheels are on dump trucks that are used in open-cut mines. These wheels are 2.6 metres (8 ft 6 in) high. The trucks can carry 170 tonnes (167.3 tons) of rock.

The smallest wheels

If you could see inside a watch, you would see what makes it work. There are lots of tiny toothed wheels called gears. The smallest is only about half a millimetre across.

Did you know?

Skateboard wheels don't have tyres. They are made of solid polyurethane (say *poll-ee-you-re-thain*). Polyurethane is a type of plastic.

A wheel journey

The long way home.

Rick Hansen became a **paraplegic** (say *para-plee-jick*) when he was 15 years old. Since then, he has used a wheelchair.

Rick became an elite wheelchair sportsperson. He won 19 international wheelchair **marathons**, including three world championships. He also played wheelchair basketball and volleyball.

Rick set the record for the longest wheelchair marathon when he travelled around the world in his chair. That's 40,075 km (24,901 miles)!

It took over two years to finish this amazing journey. Rick and his team rolled out of Vancouver, Canada, on 21 March 1985. They returned to Vancouver on 22 May 1987.

Rick had wheeled his way through four **continents** and 34 countries. He wore out 94 gloves and changed 126 flat tyres on his chair. His trip took him to Canada, the USA, the UK, Europe, Australia, New Zealand, Japan, Korea, Hong Kong and China.

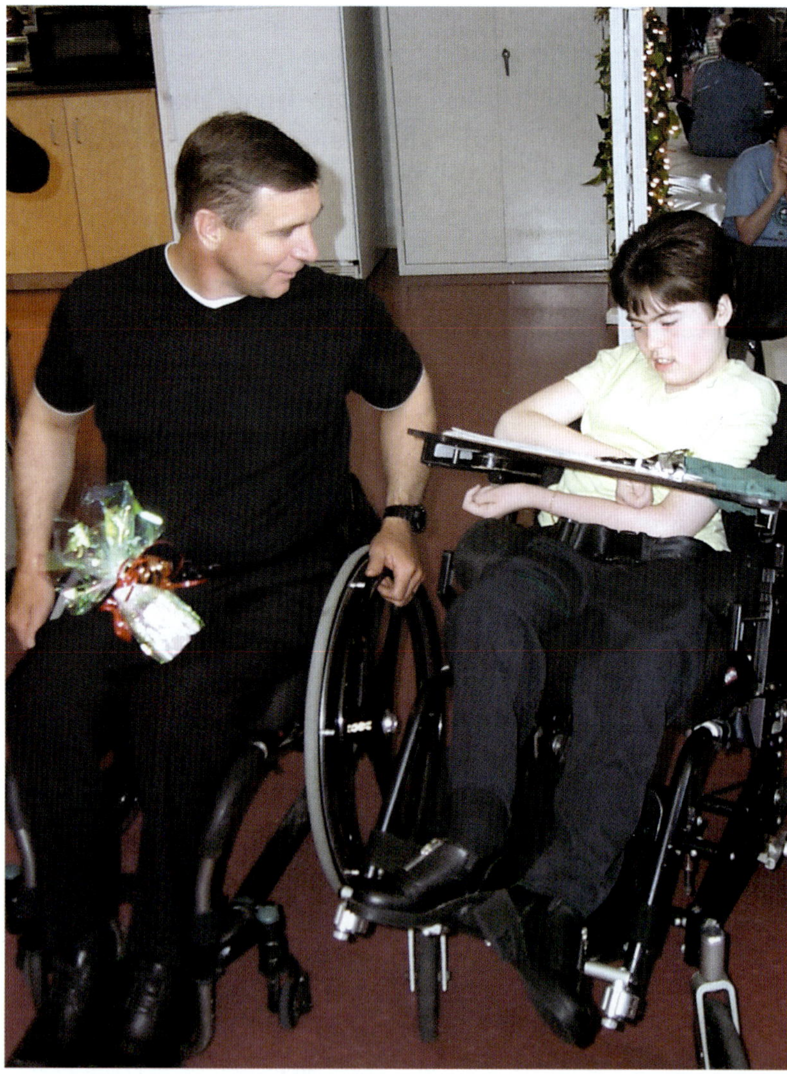

Rick Hansen heads the team at the Rick Hansen Institute to help find a cure for spinal cord injury. Here, Rick is pictured with a young girl at the opening of Rick Hansen Secondary School in Mississauga, Ontario, Canada. This school has been designed for easy access for students with wheelchairs and other physical disabilities.

A comfortable ride

Tabatha, a time-travelling journalist, zipped back to 1887. She interviewed Scottish vet, John Boyd Dunlop, who had just invented the bike tyre.

You're a vet, John. What got you interested in tyres?
A bruised bottom every time I cycled to visit sick animals.

What sort of tyres did your bike have?
Solid rubber ones. Bikes were called boneshakers.

Sounds uncomfortable.
It was. I tried tying a cushion to the bike seat. Then I had the idea that the wheels needed a cushion of air to ride on.

And how did you do that?
At first I tried wrapping pieces of garden hose around the wheels of my son's tricycle.

Did that work?
No. As soon as he sat on the trike, the tyres went flat. And my wife wasn't very happy that I chopped up the garden hose, either.

So what did you do?
I bought a new hose.

I mean about the tyres.
I filled them with **compressed** air. The air was squashed in so tightly, the tyres didn't go flat.

It was a great idea, John.
Thanks. It certainly made my life easier.

You can't just blow air into a tyre. A pump is needed to force the air in.

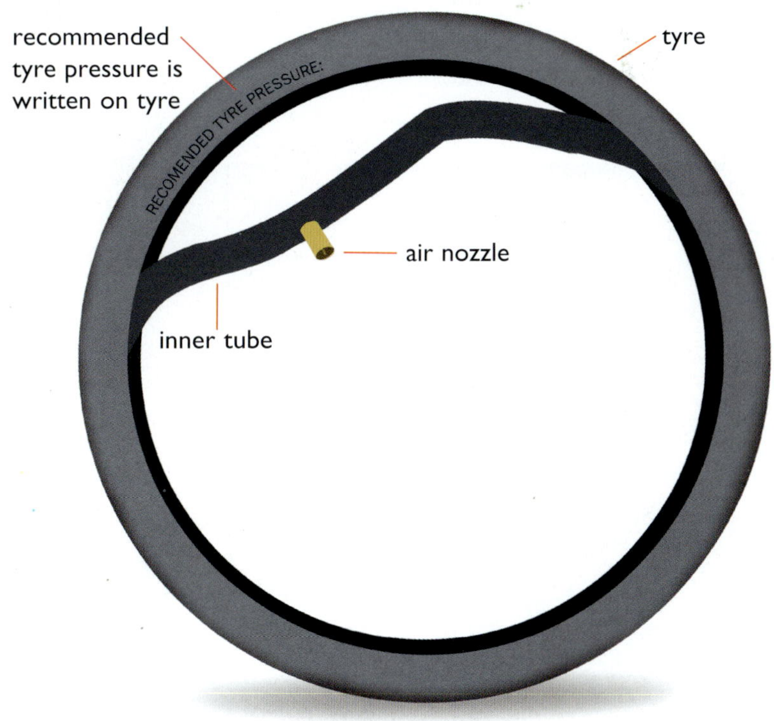

Tyres are fitted around wheels. Inside tyres, there are inner tubes filled with air. The inner tube acts like a cushion of air around the wheel.

Wheels that go round

But don't touch the ground.

WHEELS ARE used in many ways. There are even unseen wheels inside different machines.

Gears

Gears are flat wheels with teeth cut around the edge. As one wheel turns, its teeth fit into the teeth of another wheel and it moves as well. The gears turn around and around. The energy passed on by this movement is used to make machines work.

Gears are used to operate many different machines such as car engines, CD players and clocks.

Pulleys

A pulley is a simple machine used to lift loads. A rope is looped over a wheel inside the pulley. When you pull on the rope, it runs over the wheel, and lifts the load. This makes it easier to lift a heavy load.

You can find pulleys in many places such as building sites, tow trucks, rescue equipment and mountain climbing equipment.

Spinning wheel

Spinning wheels are used to make yarn or thread. Clumps of natural fibres such as goats' hair, cotton and sheep's wool are spun into long pieces of thread which can then be used to make cloth or clothing.

The spinning wheel has a wheel that is turned by hand or by a foot pedal. A rope around the wheel is attached to a spindle (say *spin-dl*).

The spindle is a round rod that is narrow at each end. As the wheel turns, so does the spindle. The person operating the spinning wheel feeds the fibre onto the spindle. As the spindle turns it stretches and twists the fibre into a thread.

Clumps of wool can be spun into thread which is used to knit clothes.

Speed check

How fast does it go?

person walking	2.4 km/h	1.5 mph
horse and cart	3.2 km/h	2 mph
bicycle	25 km/h	15.5 mph
car driving in city	60 km/h	37 mph
cheetah (fastest animal)	110 km/h	68 mph
modern high-speed train	300 km/h	186 mph
racing car	320 km/h	200 mph

Did you know?

The world record bicycle speed is 268 kilometres (166.4 miles) per hour. It was set in 1995 by Dutch cyclist Fred Rompelberg. At 50 years old, he was the world's oldest professional cyclist.

Wheel make you laugh

Roll with the punchlines.

A wheel's lament

I'm round, I roll, I run your car,
Without me, you won't get far.
You ride on me to get to places,
I bring the mail, I win the races.

All your family, all your mates,
They've all got cars or in-line skates.
On roads and rails I'm always busy,
Turning, turning till I'm dizzy.

Look under skateboards, what do you see?
No engine there, just four of me.
You drive, you ride, you have a ball,
But never think about me at all.

I roll by night, I spin by day,
Does anybody notice? No way!
I'll go on strike, one day I will,
And if I don't work you'll all stand still.

a poem by Weal Lee Grate

JOKES

Q. What did the wheels say to the boy when they fell off his skateboard and rolled down the hill?

A. *Wheel be seeing you.*

Q. What part of a car causes the most accidents?

A. *The nut behind the wheel.*

Young man: I'm stronger than you are, old man.
Old man: I bet I can wheel something across the road in a wheelbarrow that you won't be able to wheel back.
Young man: I bet you can't.
Old man: Okay, get into the wheelbarrow.

Wheels through time

What happened when.

A WHEEL TIMELINE

- 3500 BC Potters' wheels were the first wheels ever.
- 3200 BC Simple wheels were invented.
- 2000 BC Spoked wheels were invented.
- AD 1765 James Watt invented the steam engine.
- AD 1829 Stephenson's *Rocket* built. This was one of the first railway engines.
- AD 1880s The Safety bicycle invented. It was not much different to today's bicycles.
- AD 1881 Solid rubber tyres used on hansom cabs in London.
- AD 1885 First car built by Carl Benz in Germany.
- AD 1887 John Boyd Dunlop invented inflated tyres.
- AD 1955 First solar car made in Chicago, USA.
- AD 1960 Skateboards invented by bored Californian surfers when there were no waves.
- AD 1971 Wheels reached the moon on the Lunar Roving Vehicle.

The Lunar Roving Vehicle, used on Apollo missions 15, 16 and 17, allowed two astronauts to travel around the moon in order to study the geology of a particular area.

Editorial

SOME people think it would be better if wheels had never been invented. Cars and trucks can cause accidents and **pollution**. Every year, at least half a million people die in road accidents around the world. That works out to be one person every minute.

Cars and trucks also cause pollution. Vehicles that use petrol give off poisonous fumes every day. About 80% of all the **carbon monoxide** in the air comes from cars.

At *The News*, we think wheels are wonderful. People should certainly try and use cars and trucks less. But not all wheeled vehicles cause pollution. Pushbikes, scooters and rollerblades don't pollute the air.

We think someone will invent a non-polluting car soon. Then, if people drove safely, all wheels would be safer to use.

Wheels are not just used on vehicles. Imagine all the things we couldn't do if we didn't have wheels.

Letters

to the editor

Dear Editor,
I met a traveller the other day. He had ridden his pushbike from Melbourne to Darwin and back again. All that way using just two wheels and his legs! Wheels are an amazing invention.

From AMAZED AARON

Dear Editor,
I read in a history book that in 1854 the fastest coach and horses took 12 hours to travel from Melbourne to Ballarat. That's a distance of 113 km (70 miles). We drove to Ballarat last Sunday. It took us one and a quarter hours. Isn't it amazing how much faster we travel these days?

From ZOOMING ZOE

The average hamster runs up to 5 km (8 miles) per night in a wheel. No wonder this one looks tired!

Free-wheeling William
answers your questions

Dear William,
It seems that people are in love with wheels. Everybody in the world uses them. I tried to spend a day not using wheels and it wasn't possible. Did anyone ever think that wheels weren't wonderful?

From ROLLING ROSS

WILLIAM: There have been some people in history who weren't impressed with the wheel. Egyptians didn't use wheeled vehicles until 2,000 years after they were invented. For transport, they used boats on the Nile river and carted things across land on sleds.

Fact of the day

The world land speed record is 649 kilometres (403 miles) per hour. It was set in 1964 by Donald Campbell. He drove a jet-propelled car called Bluebird on a salt lake in central Australia.

Dear William,
What happens to tyres when they wear out?

From CURIOUS CHRIS

WILLIAM: Worn-out tyres are a big problem. They cannot be recycled like paper, glass and aluminium. Old tyres must be buried in the ground. This is not good for the environment. Let's hope someone will invent a new material for making tyres, or a way to re-use old rubber, soon.

What can we do with old tyres?

Dear William,
I ride along a bike path to school each day. Sometimes I see other bike riders on the road. They don't always obey the road signs. Are road signs just for cars, or do bike riders have to obey them too?

From PUZZLED PAM

WILLIAM: Road signs and traffic lights are there for the safety of all road-users. It is very important that everyone obeys the road signs. Road signs and traffic lights are for cars, trucks, buses, trams, bikes, scooters, rollerbladers and pedestrians. So if you are on your bike and you see a stop sign, make sure you stop.

Further resources

BOOKS

If you are interested in finding out more about wheels, your school and local libraries are sure to have plenty of books on all kinds of wheels.

Think about your favourite wheeled vehicles. Do you like pushbikes, motorbikes, cars, scooters, rollerblades, carts, skateboards, trains, buses, trucks, tractors, trams or aeroplanes?

In the library, look for books about your favourite wheels or wheeled vehicles in the reference section.

MUSEUM

There are many museums that have exhibitions and plenty of information about the history of wheels, bicycles, trains, planes and all sorts of automobiles. Look out for them next time you visit.

INTERNET

There are lots of great web sites about wheels. It can be hard to decide where to start looking.

Use search engines to find out more about your favourite wheels or wheeled vehicles. If you want to find out how or when they were invented, include "invention", "museum" or "history" as words in your search.

There is a terrific "history of invention" page on the cbc 4 kids web site. It has an inventions timeline and you can find out more about things with wheels such as the automobile, the bicycle, the pottery wheel, the watch and the sewing machine. Check it out at:
www.cbc4kids.ca/general/the-lab/history-of-invention/

One wheel is enough.

Glossary

ANCESTORS—people related to you who lived a long time ago.
CARBON MONOXIDE—a poisonous gas. It doesn't smell and is colourless.
COMPRESSED—something that's pressed together and forced into a smaller space.
CONTINENTS—the seven main landmasses of the world.
DEPOSITS—a layer which collects in a particular place.
FORCE—strength or power.
FRICTION—the rubbing of one surface against another.
GEAR—a toothed wheel that connects to other toothed wheels. As they turn, they change the motion of a machine.
MARATHON—a long-distance race.
NITROGEN—a gas with no colour or smell. It makes up four-fifths of the Earth's atmosphere.
PARAPLEGIC—(say *para-plee-jick)* someone who has lost the use of their legs.
POLLUTE—to spoil or make dirty.
PUNCTURE—to make a hole in something.
SUNDIAL—an instrument used for telling the time according to the shadow cast on its surface by the sun.
TAXI—when an aeroplane moves along the runway before taking off or after landing.

In Thailand, if you hail a taxi, you may not get a car. You may get a ride in a tuk tuk (say took took). This is a motorised vehicle with three wheels and a canopy over your head. It gets a bit wet inside when it's raining!

Index

If you don't have wheels, carrying a load can be hard work. But if you don't even have a donkey, it's harder still!